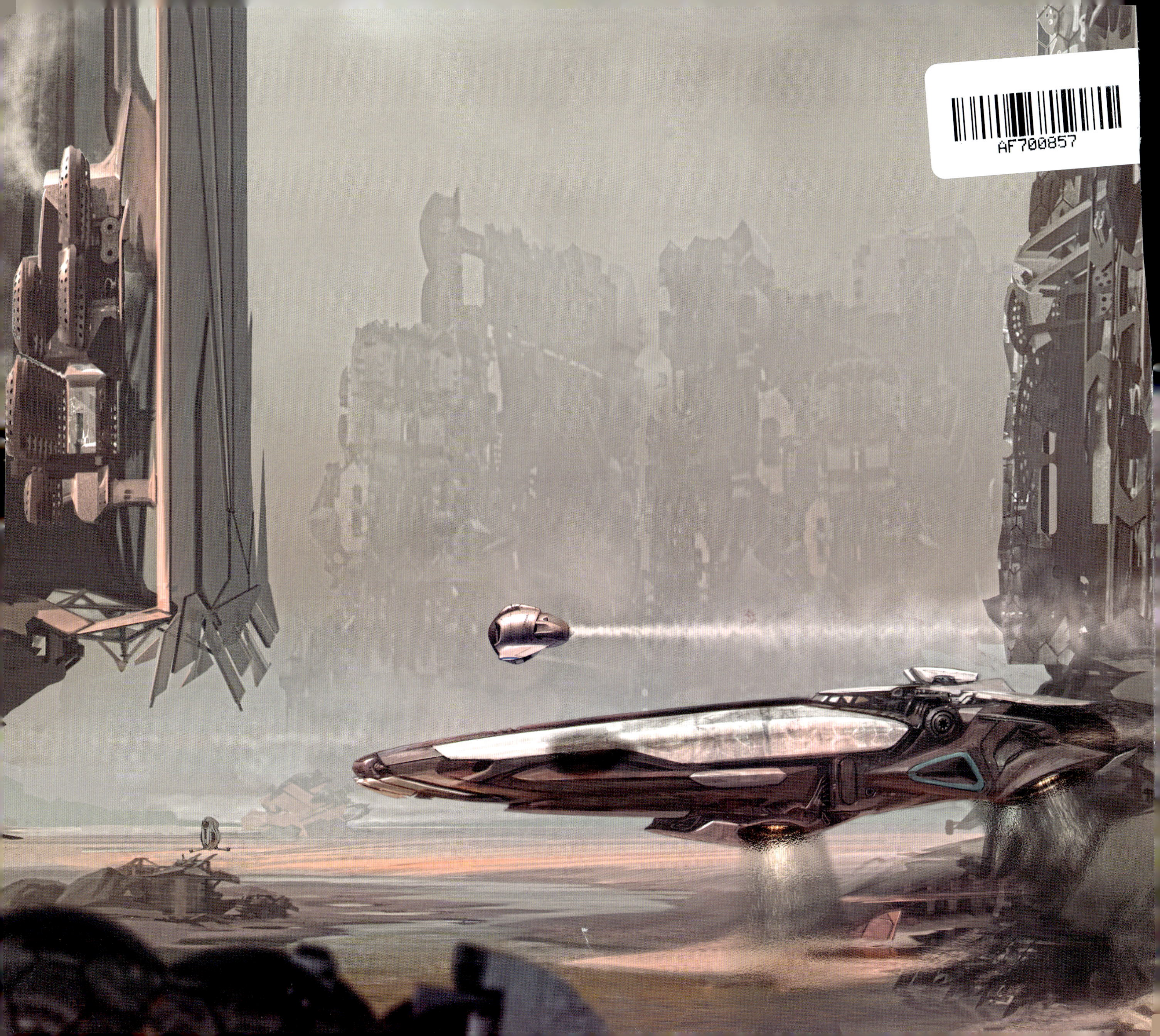

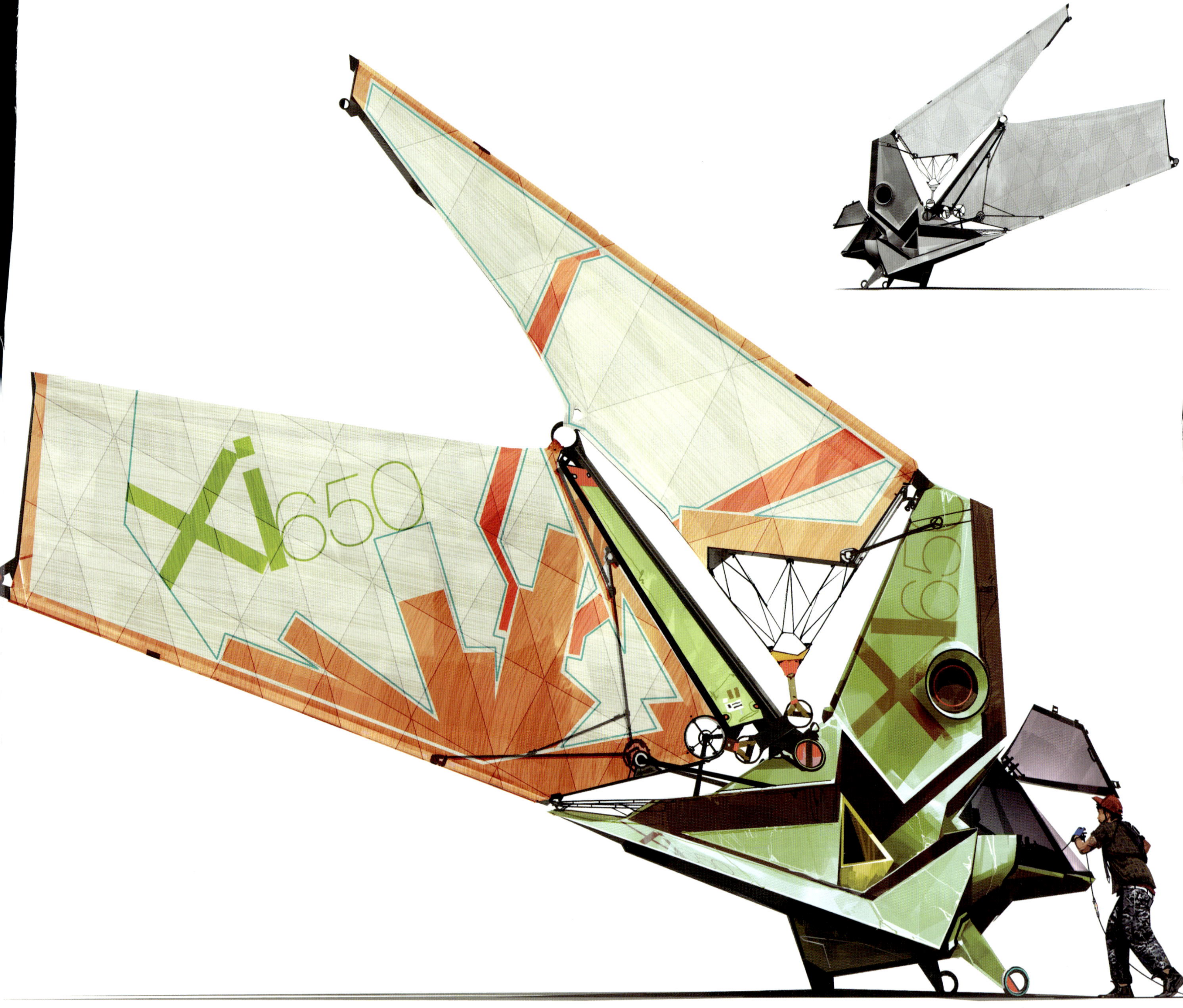
Xi650

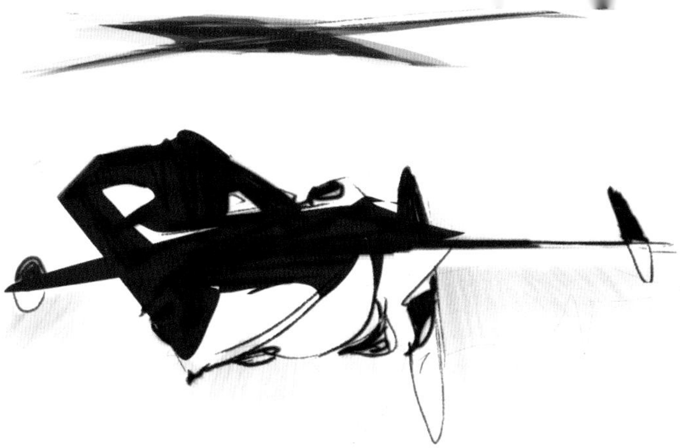

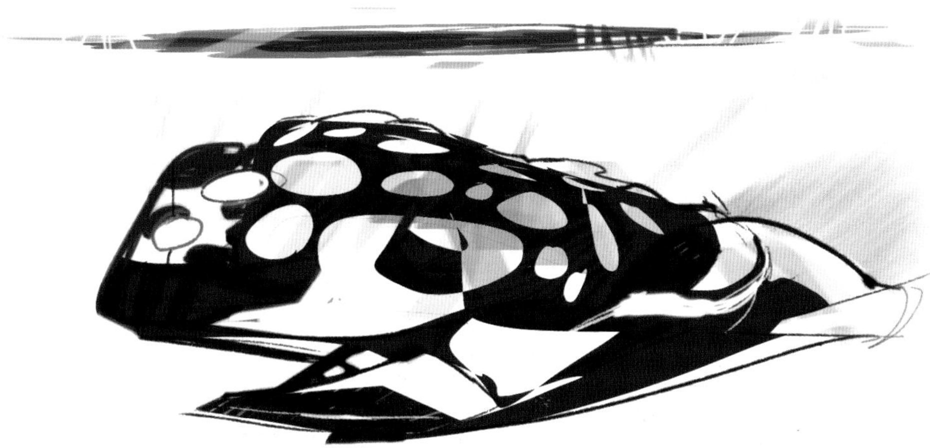

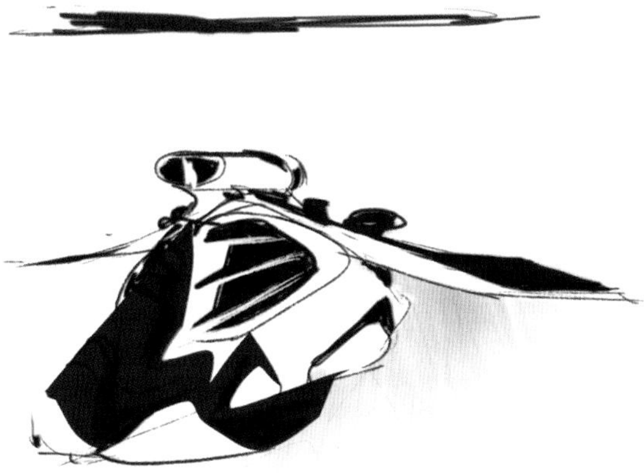

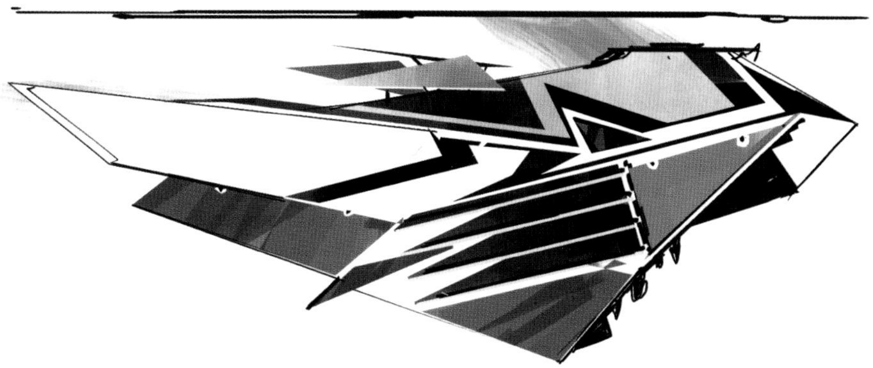

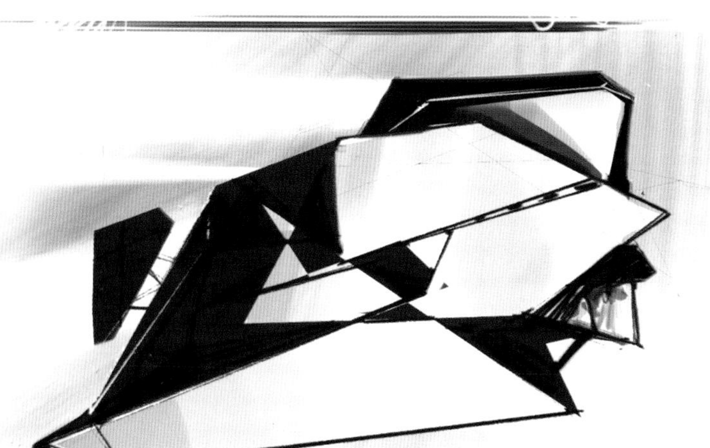

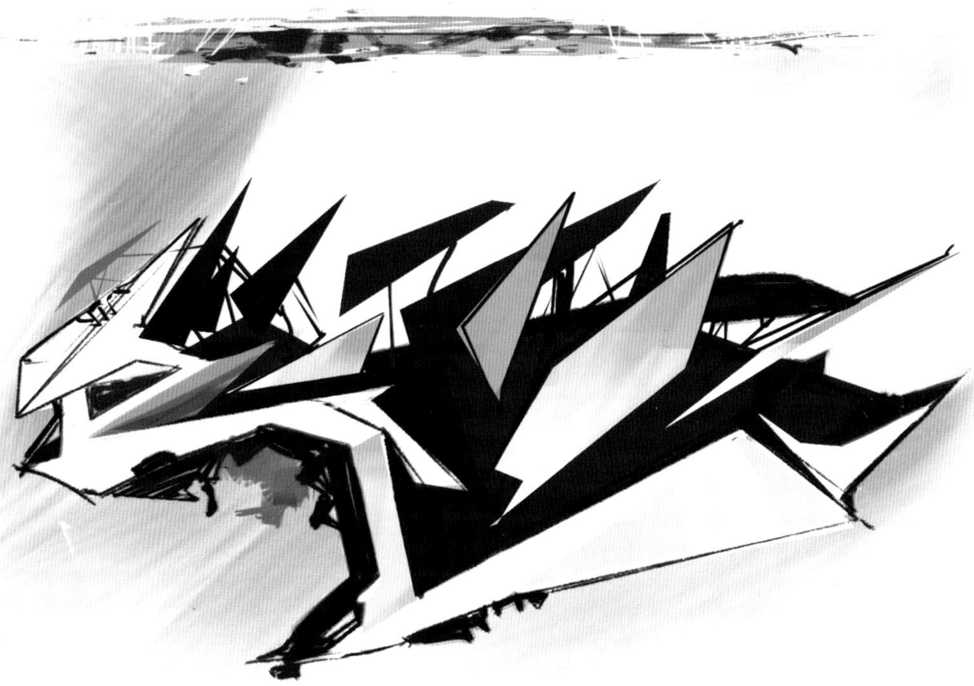

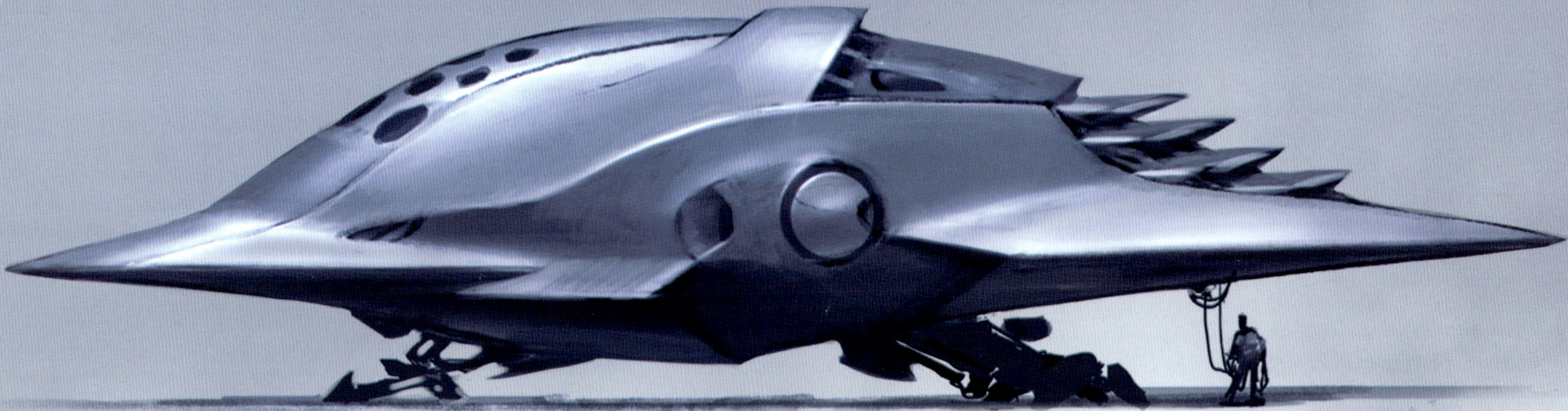

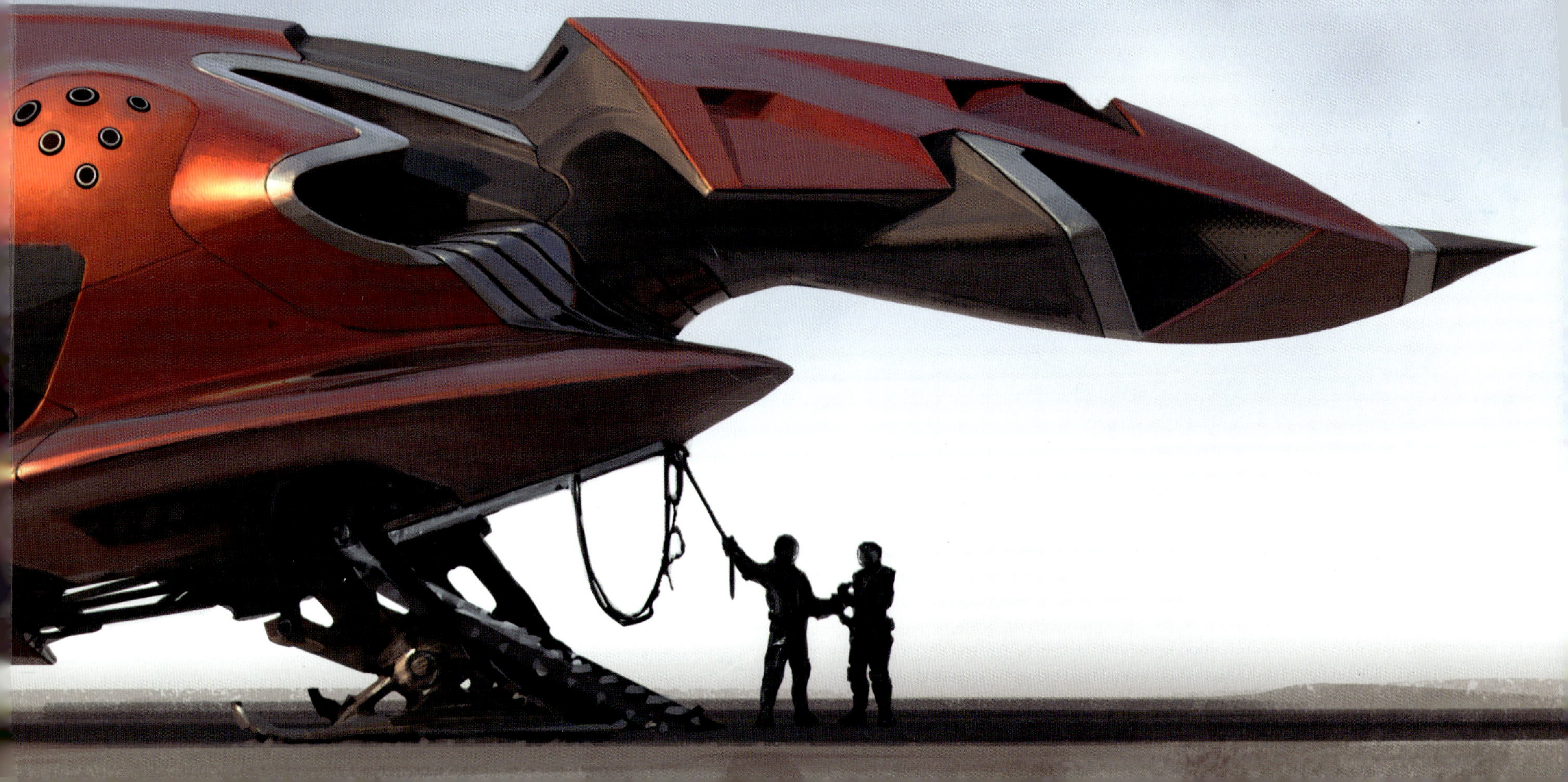

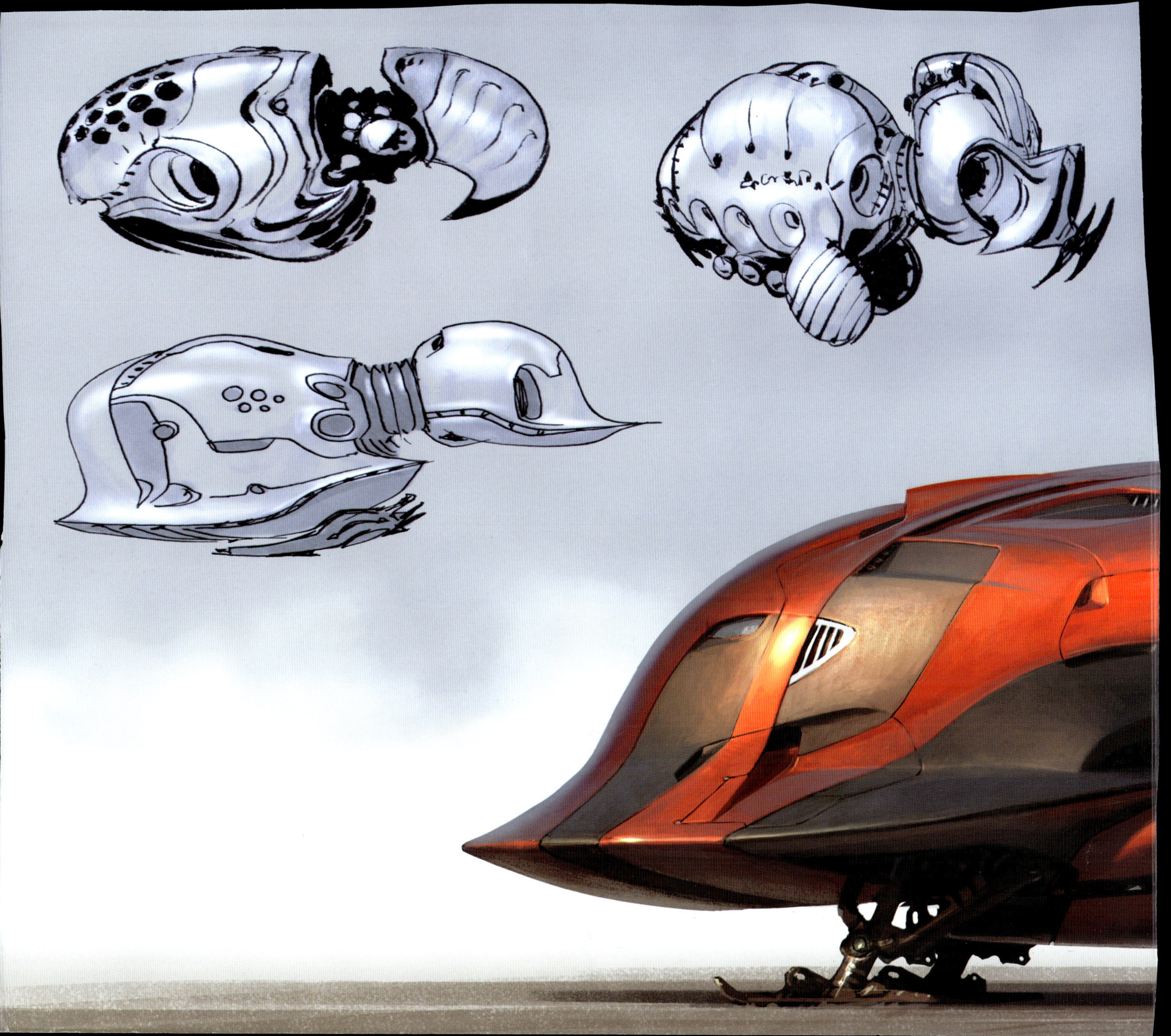

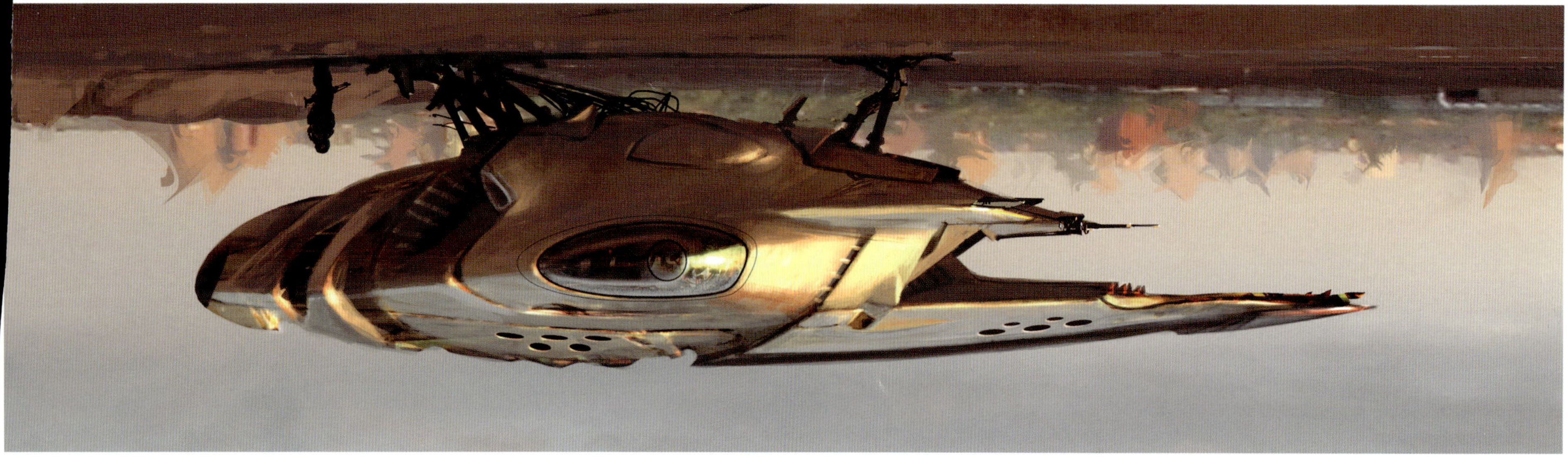

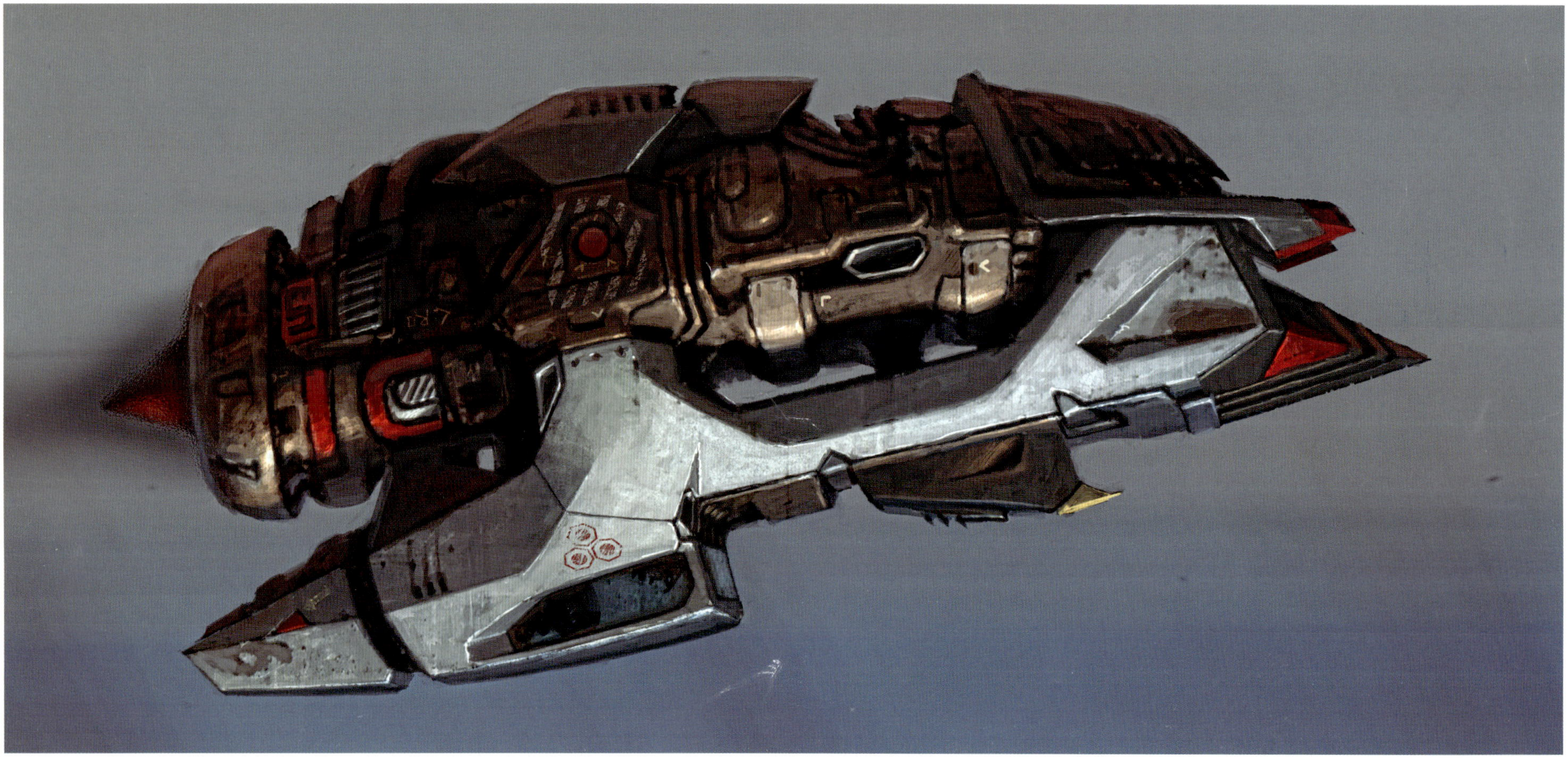

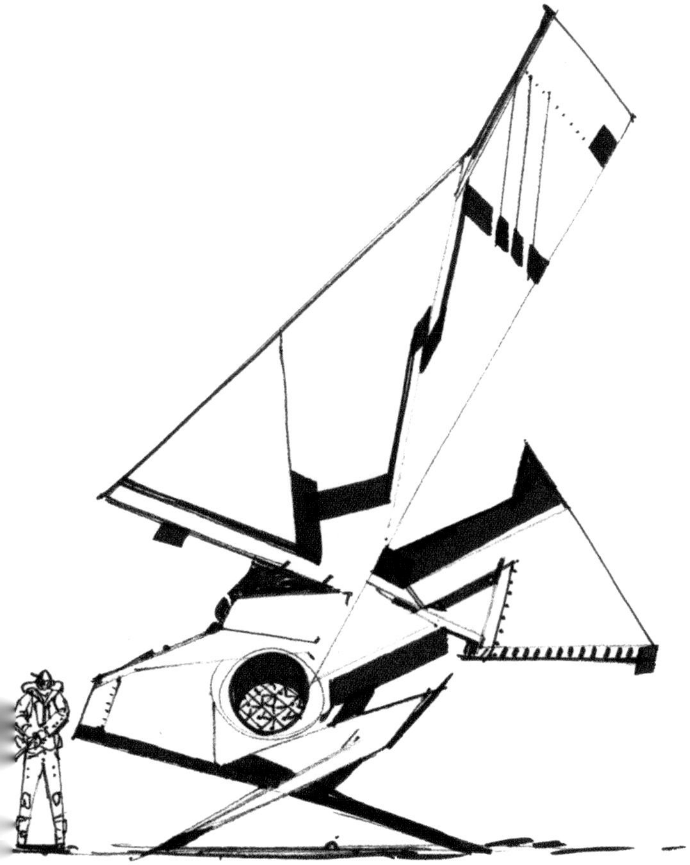

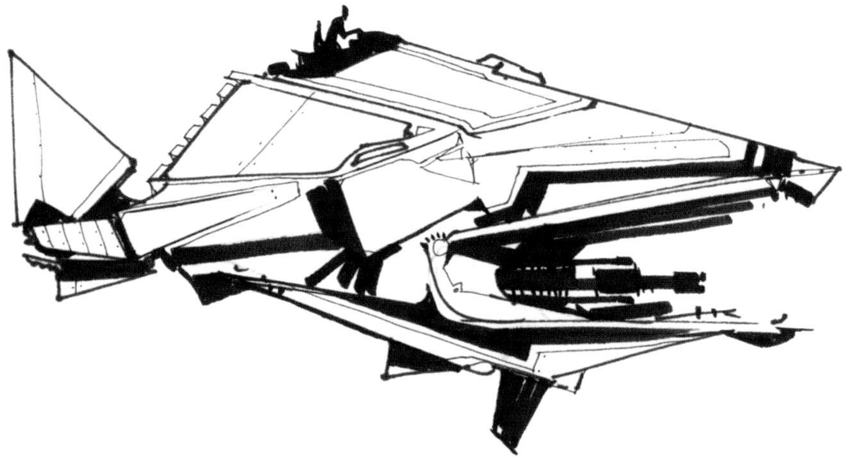

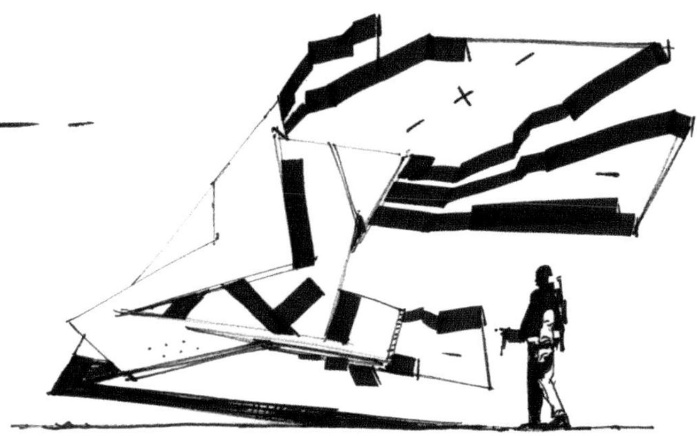

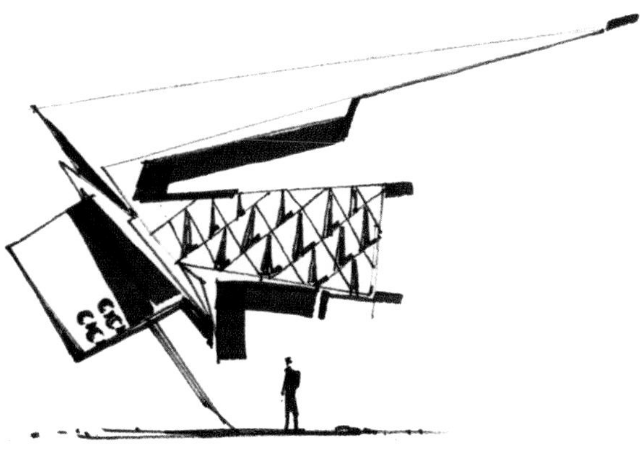

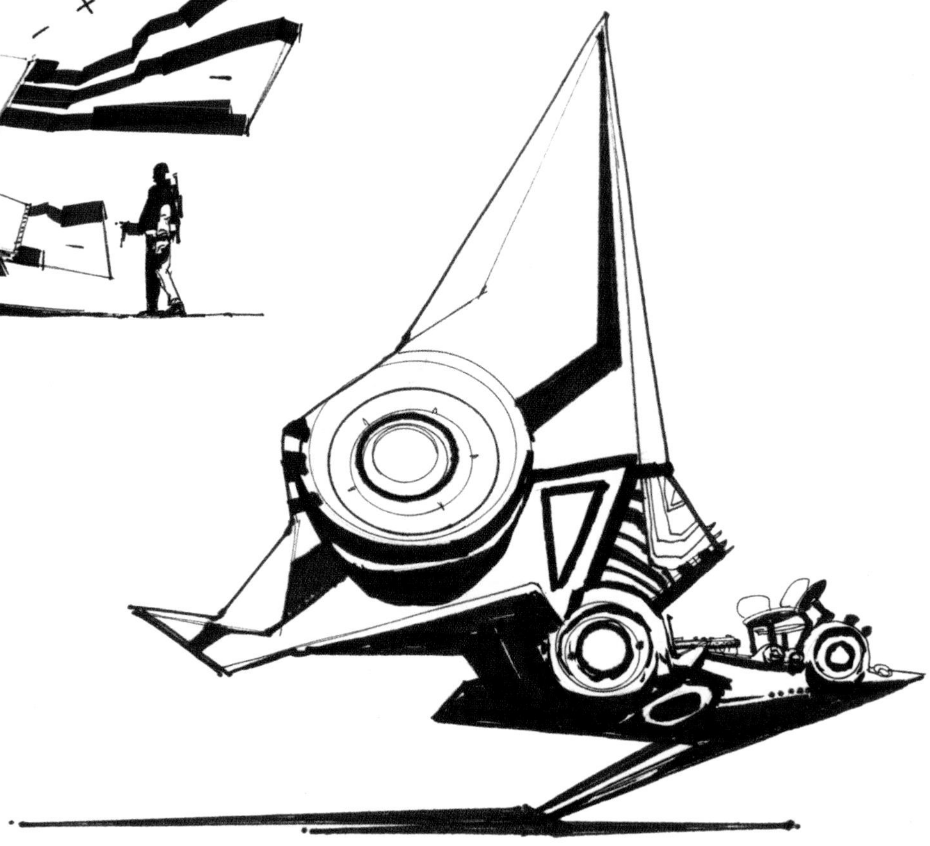

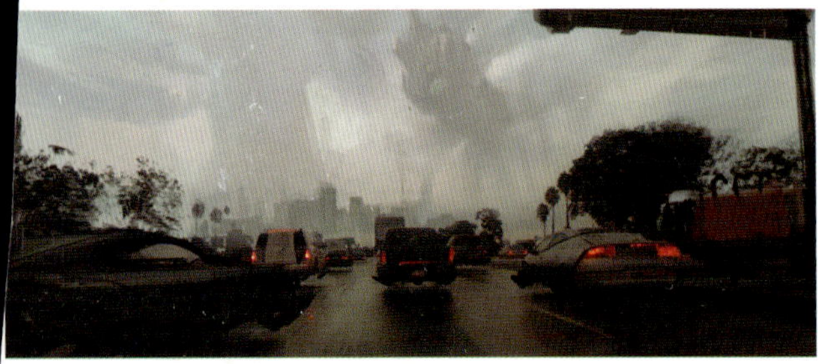

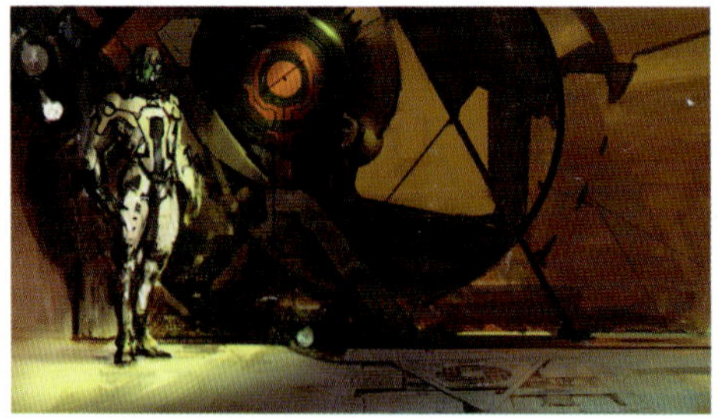

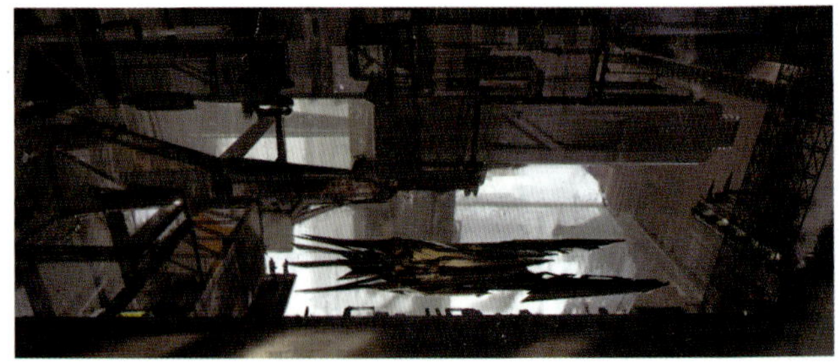

Following on the fast and successful treads of DRIVE, vehicle sketches and renderings, BLAST takes the readers on a visual development journey of spaceships. The formula from DRIVE is applied again to BLAST, with each of the three artists taking spaceships from preliminary concept sketch to final, fully rendered, photorealistic detail.

Take off with these visionaries as they launch us out of this world with their captivating spaceship creations.

**Ship Date :** 02/2012

**Pub Date :** 03/2012

**Pages :** 128

**Format :** Trade Paper

**Trim :** 9″ x 11″

**EAN/isbn :** 978-1-933492-54-4

**Price :** $24.95 USD / $24.95 CAD



ISBN
9 781933 492544

designstudio PRESS